For the animals that helped raise me. We will never forget you!

Cowpokes Who Cry LLC
Pueblo, Colorado

ISBN Paperback 979-8-9875218-6-1
ISBN Hardback 979-8-9875218-7-8
ISBN Ebook 979-8-9875218-8-5

Written by Kelcie Martin
www.cowpokeswhocry.com

Illustrated by Krystal Kramer
www.CoastToCoastArtist.com

Little Cowpoke
and
Her Ranch Friends

Written By: Kelcie Martin
Pictures By: Krystal Kramer

Little Cowpoke,
Little Cowpoke,
who are these friends I see?

Where did you find such creatures
who show love and loyalty?

We all live on a ranch,
that's the way it goes.

We stick together through the highs
and even through the lows.

The donkey,
well, he likes to bray.

He keeps our animals
from becoming prey.

Mischievous goats,
they love to play.

They provide milk, meat,
and keep the weeds away.

Little Cowpoke,
Little Cowpoke,
I'm starting to understand.

These animals do more
than just live on the land.

My horse and I guide cattle
to pastures that are new.

But whenever they see us,
they all let out a MOO!

Sometimes the herd
gets a little out of hand.

So my dog rounds them up
and listens for my command.

Sometimes we gallop fast,
and sometimes we trot slow.

We never ride in front of another cowpoke, though!

Little Cowpoke,
Little Cowpoke,
what about the cattle?

What is the reason
for riding in the saddle?

We're more than just stewards of the land,
watching over the cattle with a helping hand.

We care for them kindly,
as all cowpokes do.
Their meat feeds us well,
and many others too!

We all do our part,
caring for each other dearly.

Everyone has a purpose.
Now, can you see that clearly?

Kelcie Martin is a former cowpoke from eastern Colorado. She grew up on a 40,000 acre cattle ranch, and decided it was time to share her stories with the world! She lives in southern Colorado with her husband and twin boys. They all keep her busy while she works from home as a high school science teacher. You can find her on TikTok, Instagram, and Facebook @cowpokeswhocry.

Milton Keynes UK
Ingram Content Group UK Ltd.
UKHW051817020824
446306UK00004BA/30
* 9 7 9 8 9 8 7 5 2 1 8 7 8 *